Sally Ride Science

Now I Get It!

Adaptations

Life's Survival Strategies

Rebecca L. Johnson

Sally Ride, Ph.D., President and Chief Executive Officer; Tam O'Shaughnessy, Ph.D., Chief Operating Officer and Executive Vice President; Brenda Wilson, Vice President; Stacey Klaman, Director of Publishing; Erin Hunter, Science Illustrator; Monnee Tong, Picture Editor

Program Developer, Kate Boehm Jerome
Program Design, Steve Curtis Design Inc.
www.SCDchicago.com

Sally Ride Science
9191 Towne Centre Drive
Suite L101
San Diego, CA 92122

ISBN: 978-1-933798-58-5

Printed in the United States of America
10 9 8 7 6 5 4 3 2 1
First Edition

Cover: Feathery gills on this young Great Crested Newt larva help it breathe underwater.

Title page: Stripes break up the outline of a zebra's body, making it difficult for predators to pick out just one zebra in a herd.

Right: Giant water lilies have large, flat leaves that float on water where they can soak up the Sun.

Contents

Introduction

    In Your World 4

Chapter 1 Adapted for Life

    Adaptations Everywhere 6

Chapter 2 Natural Selection

    How New Adaptations Arise...................12

Chapter 3 Adaptations and Interactions

    From Survival to Symbiosis...................18

Thinking Like a Scientist

    Inferring and Applying...........................24

    Interpreting Data 25

How Do We Know?

    The Issue

       Vi-Vi-Vi Vibrations................................26

    The Expert

       Caitlin O'Connell-Rodwell...................27

    In the Field..28

    Technology...29

Study Guide

    Hey, I Know That! 30

Glossary 31

Index 32

Introduction

In Your World

4

The Southern Ocean surrounding Antarctica is the coldest on Earth. The temperature of the seawater is close to freezing year round. Dive in there, and you wouldn't last more than a minute or two. Why? Your body loses heat much faster in water than in air. The colder the water, the faster the heat goes. But Antarctic seals and penguins swim in the icy waters for hours. In or out of the water, they never seem to mind the cold.

What do they have that you don't? Blubber! Blubber is thick fat just under the skin. It locks in body heat—just like a warm coat that's worn on the inside. It's just one of many **adaptations** that Antarctic animals have for living where they do. An adaptation is something an organism has or does that helps it survive in a particular place.

All living things have adaptations—including you! Adaptations are the secret to surviving here on Earth.

Adaptations Everywhere

▲ **What adaptations do these goldfish have for life in the water?**

How many different kinds of living things surround you right now? People—no matter how different from each other— count as one kind, or **species**. Is there a dog snoozing at your feet? Goldfish in an aquarium? Grab a pen and start a list!

A list of all Earth's species would be incredibly long. Scientists have described about 1.8 million species so far. They think there are millions more. Each and every one of those species has unique adaptations. Adaptations help living things survive in whatever **ecosystem** they call home.

An aquarium is a mini-ecosystem, a tiny version of a lake. All the things living in it have adaptations for life in water. Gills allow fish to breathe underwater. A snail's hard shell protects it from hungry fish. Tiny pockets of air in the leaves and stems of water plants keep them standing tall or even floating near the water's surface.

Everything in Its Place

If you woke up and saw camels grazing on your front lawn, you might think you were still dreaming. Just as fish belong in water, camels belong in deserts. That's because camels have adaptations for living in hot, dry ecosystems.

Species are different from place to place because ecosystems are different. That's why you find certain kinds of living things in one area but not another. With thick fur and snowshoe paws, polar bears are adapted for life in Arctic ecosystems. Web-footed frogs are adapted for life in the shallow waters of wetlands. Many rainforest plants have leaves with pointed, "drip tips" for quickly shedding water. They are adapted for life where it rains almost every day.

Despite their differences, living things everywhere have the same basic needs. They must have water and food to survive. Adaptations help them get these essentials, no matter where they live.

Beetle Juice

Beetles in Africa's Namib Desert have an adaptation for getting water. When fog blows in from the coast, the beetles lean into the wind. Water drops collect on their bumpy backs and trickle into their mouths!

▶ Sandstorm? No problem! A camel can close its nostrils to keep sand from blowing up its nose.

The Bottom Line | Every ecosystem is home to living things with adaptations that help them survive in that place.

Then Along Comes Change

Change is hard. Ask anyone who has ever started at a new school or moved to a different town.

Change is hard for every living thing. When an ecosystem changes, all the species that live there are affected. Even small changes—like a dry spell or a long winter—can make life tougher.

What if changes to an ecosystem are large or sudden, or last a long time? Adaptations may no longer help species get what they need. When survival tools stop working, living things die.

When all the members of a species die, the species becomes **extinct**. Take dinosaurs. For more than 150 million years, dinosaurs ruled Earth. Then about 65 million years ago, they vanished—they went extinct. Why? It's because a giant asteroid—15 kilometers (9 miles) wide—crashed into the Gulf of Mexico. Dust and dirt thrown into the air by the impact circled the globe. This triggered a global winter. Ecosystems worldwide experienced a long, chilly twilight. Creatures that couldn't adapt to the harsh conditions died out. That's what happened to the dinosaurs along with more than 50 percent of all species living on Earth at that time.

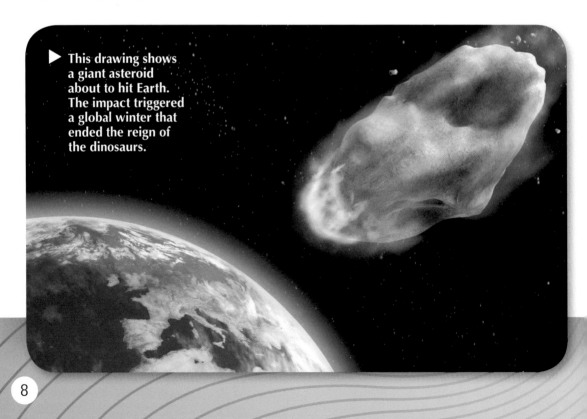

▶ This drawing shows a giant asteroid about to hit Earth. The impact triggered a global winter that ended the reign of the dinosaurs.

Proof of the Past

What proof do we have that dinosaurs once roamed Earth? Or that ancient forests were made up of now-extinct trees? **Fossils**! Fossils are the remains or traces of living things that existed long ago. They range from bones and teeth to footprints and impressions left by feathers, fronds, and scales.

▶ Sudden climate changes spelled extinction for the dinosaurs.

Using fossils, scientists have pieced together a history of life on Earth. It shows that extinctions are nothing new. They're not even uncommon. Species have been going extinct ever since life first arose about 4 billion years ago. In fact, 99.9 percent of all the species that have ever lived on Earth no longer exist.

But don't get the idea that everything dies out. Species can and do survive when ecosystems change. They survive by changing, too.

▲ Dinosaurs might have rested in the shade under giant tree ferns. Fossils of tree ferns provide clues about plant life millions of years ago.

The Bottom Line | **When ecosystems change, living things struggle to survive—and sometimes go extinct.**

Evolution at Work

That's right—species also change. Not overnight, but gradually. This slow change in living things is called **evolution**. Evolution produces new adaptations in living things over long periods.

Keep in mind that entire species evolve, not individuals. A single organism can't adapt to a change in its surroundings by altering its body. An animal, for instance, can't suddenly grow long fur or develop a layer of blubber to stay warm in an ecosystem that's getting colder. If the cold affects a food the animal eats, the animal might switch to eating something else. But one animal's change in diet isn't the same as an adaptation that develops through evolution.

The evolution of new adaptations can take millions of years. And it always happens to an entire **population** of living things, not just one.

Evidence Left BeHIND

Fifty million years ago, ancestors of whales had legs and lived on land. These ancient species gradually evolved flippers and other adaptations that made life in the ocean possible. Some modern whales still have small bones for hind—back—legs. The bones are "leftovers" for legs whales no longer have.

◀ Like seals and penguins, whales have thick blubber to stay warm in cold water. It's an adaptation that took millions of years to evolve.

Endless Adaptations

Meat-eating plants? Fish that change sex? No, these aren't strange sci-fi characters from another solar system. They are Earth dwellers, just like you. Their particular adaptations might seem bizarre at first glance. But they are tools for survival, as all adaptations are.

▲ **If the only male in a family of parrotfish dies, one of the females will change sex to take his place.**

You could spend a lifetime studying adaptations in living things and you'd still just scratch the surface. Millions of different species means millions of different adaptations. There are so many, it's hard to grasp.

All these different species—and their adaptations—are the result of billions of years of evolutionary change. Evolution is what has led to the great **biodiversity** of life on Earth. It's produced all life on the planet, past and present. That means dinosaurs and death cap mushrooms, oak trees and okapis, *E. coli* and eels . . . and everything else.

◄ **A sundew plant uses sticky drops to catch an insect meal.**

The Bottom Line | **Evolution is the change in living things over time.**

How New Adaptations Arise

Okay, so living things evolve over time. But how does evolution work? How do species actually develop new adaptations? The answer is **natural selection**. Natural selection is the driving force behind evolution. It all starts with **traits**.

Traits are features or characteristics that living things have. You see some of your own traits every time you look in a mirror. Is your hair curly or straight? Are your eyes blue, green, or brown?

Do people say you've got your mother's nose or your father's chin? Most traits are inherited. They're passed from generation to generation.

Traits don't just happen. They're controlled by **genes**. Genes are like little sets of instructions all packaged together in the DNA inside your cells. Genes control almost every trait you have. Genes in the DNA of other living things control their traits, too.

◀ **Your traits are unique to you.**

▲ Look closely. The stripes on each zebra are slightly different.

Variations, Invariably!

Have you ever tried one of those puzzles where you have to find the two shapes or patterns that are exactly the same? If you played that game with living things, you'd get pretty frustrated. That's because no two living things are exactly the same.

Take zebras. In any population of zebras, no two are identical. Each one has slight differences, or variations, in its traits. One zebra might have darker stripes. Another might have longer legs.

Still another might have bigger eyes. You get the idea.

What's true for a population of zebras is true for any population of living things—from paramecia to pine trees and from parrotfish to people. Individuals in a population—even very closely related ones—have slight variations in their traits. That's because they all have slight variations in their genes. And genes, as you know, control traits. So when genes change, what happens? You guessed it—traits change, too!

The Bottom Line | **Members of any population of living things have slight variations in their genes and, therefore, in their traits.**

▶ **Now you see it . . .**

Natural Selection in Action

So what do variations in genes and traits have to do with natural selection? Everything! Here's how it works.

Imagine a population of tortoises living on an island. Most of the tortoises can pull their legs and heads only partway into their shells. A few tortoises, however, have a variation of this trait. They can pull their heads and legs all the way inside their shells.

For a long time, this variation doesn't seem to matter much. Then rats arrive on the island. The rats attack and kill the tortoises that can only pull themselves partway into their shells. The tortoises that can pull their heads and legs all the way in, however, survive unharmed inside their shells. These tortoises have an advantage. Their variation helps them survive.

▼ **. . . now you don't! Adaptations like a tortoise's ability to withdraw into its shell are the product of natural selection.**

The pull-all-the-way-in tortoises reproduce. They pass the genes for this trait to their **offspring**. Offspring with the trait also survive. They reproduce and pass the genes to their offspring, too. After many generations, all the tortoises in the population will be able to withdraw completely into their shells. They will have evolved a new adaptation that helps them survive in their rat-infested ecosystem.

In time, evolution through natural selection can produce living things that are so different from their ancestors that they become a new species. Every new species fills a unique **ecological niche** in its ecosystem. Its adaptations enable the species to live in a certain place, eat certain kinds of food, and behave in certain ways. The new species fits in with the other species that call that ecosystem home.

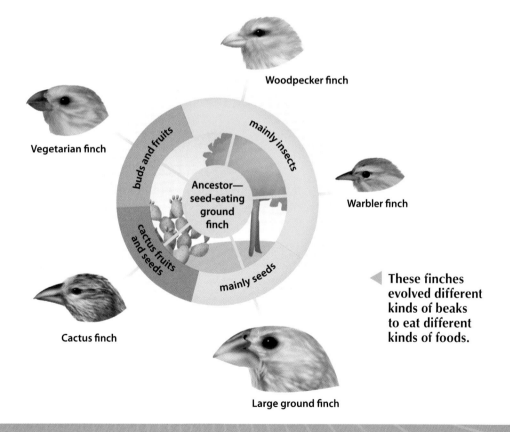

Woodpecker finch

Vegetarian finch

buds and fruits

mainly insects

Ancestor—
seed-eating
ground
finch

Warbler finch

cactus fruits
and seeds

mainly seeds

Cactus finch

Large ground finch

◄ These finches evolved different kinds of beaks to eat different kinds of foods.

The Bottom Line | **Natural selection leads to the evolution of new adaptations and sometimes new species.**

Artificial Selection

Through natural selection, living things evolve in response to changes in their surroundings. But it's a very slow process—too slow to see in action.

Evidence of **artificial selection**, however, is all around you. Artificial selection is similar to natural selection. The difference is that people, not Mother Nature, decide what traits living things have. People breed animals and plants with certain characteristics to produce offspring with desirable traits.

People have been selectively breeding dogs, farm animals, flowers, and crops for thousands of years. Many of the fruits and vegetables we eat are the result of artificial selection. For example, you can thank this process for grapes, oranges, and watermelons that don't have seeds. No spitting required!

Good Dog, Better Dog

Want a dog? You've got about 400 varieties, or breeds, to choose from! These breeds are the result of people using artificial selection to produce dogs that have certain desirable traits. Different breeds come in different sizes, shapes, and coats—even personalities!

▼ Yum! A tangelo is a cross between a tangerine and a grapefruit.

Evolution of an Idea

You've probably heard of Charles Darwin. He was a naturalist back in the mid-1800s. After spending five years studying plants and animals—and their fossils—all over the world, Darwin returned to England. There he spent many more years pondering what might cause species to change over time.

Artificial selection got Darwin thinking about traits and adaptations. If people selectively bred plants or animals to produce offspring with certain traits, why couldn't nature do something similar? Darwin studied artificial selection in pigeons, rabbits, cabbages, and other types of living things. He realized that it was a speeded-up version of the process that gives rise to new species in nature.

In 1859, Darwin published a book called *On the Origin of Species*. In it, he proposed that living things evolve through natural selection. The theory of evolution by natural selection fundamentally changed people's understanding of the living world. It provides a scientific explanation for Earth's incredible biodiversity.

▲ Darwin bred pigeons, like the English Carrier (top) and the English Pouter (bottom), but not for fun. He used artificial selection to understand how natural selection works.

People use artificial selection to selectively breed living things to produce offspring with certain traits.

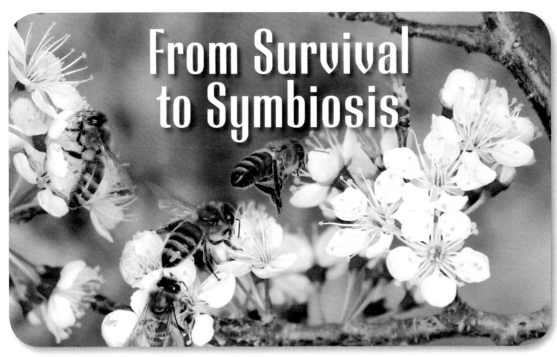

From Survival to Symbiosis

▲ **Many flowers have Velcro-like petals that make it easy for bees to grip as they sip nectar.**

Here's a challenge. Try to find a living thing in nature that's not interacting with another living thing. How about that robin on the lawn? Nope—it's tugging on a worm. How about those plum blossoms in your neighbor's garden? Hmm . . . just look at all those bees.

In every ecosystem, populations of different species form a complex **community** of living things. Those populations are constantly interacting with each other. They're also interacting with their surroundings. Robins take in oxygen from the air. Their cells use it during cellular respiration. Sunflowers take in carbon dioxide from the air. Their cells use it during photosynthesis. No matter what the interaction, though, adaptations are involved

Take ecosystems where resources are in short supply. You'll find living things with adaptations interacting in ways that help them make the most of what's there. Many desert plants, for example, have roots that spread far and wide. When rain falls, the roots are ready to soak up every drop of water.

▲ Beware, all mice! Using the heat-sensitive pits beneath its eyes, this rattlesnake can detect even faint heat coming from warm-blooded prey.

Who's for Dinner?

Getting enough food can be the difference between life and death. So it's not surprising that living things have lots of adaptations for doing just that. Many animal **predators**, for instance, have sharp vision, spectacular hearing, and a keen sense of smell. Some have adaptations that allow them to feel the slightest movement in the air or water around them.

Dagger-like claws, pointed teeth, and sticky webs are just a few adaptations predators have for catching and holding **prey**. Some, like microscopic amoebas, use their entire bodies to capture a meal!

Many predators have adaptations that help them avoid having to compete with other species for food. An owl, for example, hunts at night when hawks and eagles are asleep.

The Weird!

Lassoing Lunch

Lurking in the soil are threadlike fungi. Some of the threads grow tiny loops. The loops are traps. When a worm crawls through one, the loop tightens like a noose. Then digestion begins!

Magnification: 300x

The Bottom Line | **Living things have adaptations that help them interact with each other and their environment.**

▲ Any predator that tries to bite an agave will end up with a mouthful of spiny leaves.

▲ In winter, an Arctic hare's white fur is perfect camouflage in snowy surroundings.

Clever Defenses

Never fear—organisms that risk becoming something else's lunch aren't defenseless. Their adaptations—teeth, claws, and much more—help them survive.

Can you imagine biting into a cactus? How about a porcupine? Sharp spines are an excellent defense. Tough scales, a coating of slime, and poison parts are also adaptations that keep even the hungriest predators away.

It's possible to escape being munched on in other ways. Some organisms have evolved to look like stuff that's not fit to eat. Certain caterpillars, for example, are dead ringers for bird droppings. Other insects are the spitting image of old leaves or twigs. Many animals have adaptations that allow them to blend in with their background, too. Predators can't eat what they can't see!

▶ Were you fooled? With its wings folded, this butterfly looks like a dead leaf.

When the Going Gets Tough . . .

Freezing cold or scorching heat? No problem. Droughts or daily downpours? Bring them on. Everywhere you look, you'll find living things with adaptations that help them survive environmental extremes.

In places with cold winters, some animals **hibernate**—they sleep the frigid months away. Others avoid the cold altogether. They **migrate** to warmer places. Some plants survive winter thanks to a thick skin of protective bark. Others don't bother staying warm.

They die back to the ground as temperatures drop. In spring, they send up new shoots and start growing again.

In hot places, you'll find lots of adaptations for survival too. Many desert plants have small leaves—or none at all—so they don't dry out in the heat. Many desert animals live underground during the day and only come out at night. Geysers and volcanic pools can be much hotter than deserts. Yet even in these extreme ecosystems you'll find **microbes** with adaptations for surviving the heat.

▲ A wood frog has spent the winter frozen solid on the forest floor. In spring it will thaw out and hop away.

▲ Some bacteria and archaebacteria like it hot!

The Bottom Line | Some adaptations help living things defend themselves against other living things, while others help them survive in their environment.

Working Together

Your chemistry homework is due tomorrow. But you're stuck on question 17. You call your lab partner, knowing she can help.

By teaming up with a partner, you improve your chances of surviving chemistry class. Other living things team up with partners, too. They work together with another species in ways that help them both survive. This is called a **symbiosis**. Symbioses are some of the most complex adaptations in the living world.

On tropical coral reefs, both clams and corals partner with tiny plantlike algae. The algae live, safe and sheltered, inside the animals' bodies. Using energy from sunlight, the algae make food for themselves and their animal hosts.

They also add fresh oxygen to the water. It's a nifty arrangement in which everybody wins.

So That's Why!

Global warming has got scientists pretty worried about tropical coral reefs. Why? When ocean water warms above a certain temperature, algae leave their coral hosts. Without their algae partners, corals can't survive. As the corals go, so goes the vibrant reef ecosystem.

◀ **Giant clams have Sun-loving algae in their bodies. Imagine having houseplants under your skin!**

▲ A parade of leafcutter ants heads home with its prize.

Tropical leaf-cutter ants slice and dice leaves into small pieces. They carry the pieces to their underground nests. Worker ants chew the leaves into a paste. The paste is food for a fungus, which in turn is food for the ants. The ants and the fungus depend on each other. Neither can survive alone.

Did you know that you are a partner in a symbiosis, too? About 100 trillion bacteria live in your digestive tract. In exchange for a warm and safe place to live, they manufacture vitamins and help you get the most from your food.

So you see—you're living proof! Adaptations are everywhere, part of every species. Shaped by natural selection, adaptations are the tools that make survival on Earth possible. They have since life began.

The Bottom Line | **A symbiosis is a partnership between different species that have adapted to live together.**

Insect-eating bats that live where winters are cold have a problem. Insects disappear at the end of summer. Without food until spring, how do bats survive? They hibernate!

Inferring

Hibernating bats fall into a deep sleep called torpor. In torpor, a bat's heart beats only a few times per minute. Its body temperature falls many degrees. Torpor is an adaptation for saving energy. Hibernating bats survive without eating for several months.

Bats wake from torpor if the temperature drops below 0°C (32°F). They shiver to raise their body temperature so they don't freeze. But warming up uses energy. If bats wake too often, they won't survive until spring. They'll starve to death. Knowing this, scientists infer that bats hibernate in places with a fairly constant temperature. Where are such places? In caves!

Applying

Many species of bats are endangered. One reason is that people love to explore caves. But these cavers sometimes disturb hibernating bats by opening up a new entrance. This can make a cave too cold in winter. Scientists apply what they know about bats to restore conditions in caves.

◀ **Frost coats the fur of this hibernating bat.**

By returning caves to their natural state, scientists are helping bats survive and their populations grow.

Interpreting Data

Scientists monitored the temperature in a cave where the bat population has been shrinking. Plot the temperature data shown below on a graph.

Date	Cave Temperature
September 15	4.44°C (40°F)
October 1	1.67°C (35°F)
October 15	1.67°C (35°F)
November 1	0.55°C (33°F)
November 15	-1.67°C (29°F)
December 1	1.67°C (35°F)
December 15	1.11°C (34°F)
January 1	-5.55°C (22°F)
January 15	1.67°C (35°F)
February 1	-1.67°C (29°F)
February 15	0.55°C (33°F)
March 1	4.44°C (40°F)
March 15	9.44°C (49°F)

Your turn! Use the information on your graph and on these pages to answer the questions.

1. What were the warmest and coldest temperatures recorded in the cave?

2. Make an inference. How many times did the bats in the cave wake up during the winter?

3. Why might the bat population be shrinking?

4. Cave explorers chipped out a new entrance to the cave last summer. How might the new opening in the cave and the temperature inside the cave be related? What could be done to protect the bats?

Vi-Vi-Vi Vibrations

On the savannahs of Africa, where lions prowl, the more warning an elephant has of danger, the better. Hearing alarm calls from other elephants miles away can give them the warning they need to survive.

Cartoons often show elephants listening with their big ears fully outstretched. But that's only one way elephants listen. Sometimes they hold their ears flat against their heads, roll forward on their tiptoes, and lay part of their trunk on the ground.

How does that help?
When elephants strike that pose, they're actually using several adaptations to sense danger. They're not trying to hear sounds traveling through the air. They're actually picking up vibrations that travel through the ground— just like the waves you would feel in an earthquake.

Both the trunk and the edges of an elephant's feet are packed with cells that are very sensitive to vibrations. These cells send signals to the brain. So, in a way, they enable elephants to listen with their feet!

Even the bones in elephants' ears help them sense vibrations. When the animals freeze in place, ground vibrations, or seismic cues, travel from the toes into the foot bones and then all the way up to the ears. The vibrations shake an enlarged middle ear bone, which sends a signal to the brain.

▶ When a distant herd of African elephants stomps and rumbles, other elephants get the message through their trunks and toes.

Caitlin O'Connell-Rodwell

Behavioral Ecologist

STANFORD UNIVERSITY
SCHOOL OF MEDICINE

▲ Caitlin's research has taken her all over the world. Here, she's leaping for joy near the summit of Haleakala Crater in Hawaii.

◀ Ever since Caitlin was a young girl, being around animals has put a big smile on her face.

For as long as Caitlin O'Connell-Rodwell can remember, she's loved watching animals. Growing up with woods and a stream nearby, she was fascinated by the salamanders, frogs, crayfish, and possums she spotted there.

But it wasn't until after college that she realized she could study wild animals for a living. A friend invited her into a field biology program where she spent the summer observing insects. From then on, she was hooked.

As a young researcher, Caitlin jumped at the opportunity to work in Namibia studying elephant behavior. Her careful observations over the course of more than a decade have shown that elephants listen with more than just their ears. And she's seen firsthand what social creatures they are.

"The more I study elephants," Caitlin says, "the more special I think they are, and the luckier I feel to be able to get to know them better."

Caitlin O'Connell-Rodwell and her research team peer out from a tower 20 feet (6.1 meters) over a water hole in southwestern Africa. *Shh.* A herd of elephants has just arrived to drink and bathe. Caitlin whispers, "Start." The team observes the elephants for five minutes and then begins their experiment.

Years earlier, Caitlin recorded the low rumbles and stomping sounds an elephant made when two hungry lions approached the water hole. She converted those sounds into signals that can be thumped out and transmitted though the ground with a device called a shaker. Would today's herd get the message?

Caitlin and her team watch the response—the elephants freeze, lean forward, turn toward the signal, and then group together. "They're responding to our seismic cues," Caitlin says.

▶ Caitlin first learned about communicating through seismic vibrations by studying insects. Careful observations led to her discovery that elephants do the same.

▲ Caitlin's team buries shakers that pound out ground signals.

The shakers Caitlin and her team use in Africa to deliver ground signals to elephants aren't fancy scientific devices. People buy the same technology every day. They use it to make their home or car music pulse to the beat. Caitlin encases the shakers and buries them underground. Wires connect them to a computer and a tape recorder.

INVENTION CONNECTION

Choose Your View

You're a field biologist in Africa. You need to design a hiding place from which your team can observe animals at a nearby water hole. Read the guidelines and then sketch your idea. Your hiding place

> must protect humans from hungry animals.
> must provide a clear view of the water hole.
> can be elevated or buried.
> must hold five researchers and their equipment.

▼ Cape buffalo might stampede if they realize someone's watching them wallow at a water hole.

Oh, By the Way

As adults, male and female elephants live separately. Caitlin has noticed that male elephants, called bulls, react less to her recorded alarm signals than the females do. That may be because young elephants are the most vulnerable targets, and the females need to react in order to protect them.

Hey, I Know THAT!

Impressed by the adaptations you've read about in this book? Now it's your turn to shine. You possess an adaptation most other living things don't. You can communicate to others what you've learned. On a sheet of paper, show what you know as you do the activities and answer these questions.

1. Make a list of five plants and animals pictured in this book. Identify at least one adaptation each has and explain how it helps that plant or animal survive. (page 5)

2. Why are the plants in a rainforest different from those in a grassland? (page 7)

3. What is the outcome of evolution? (page 10)

4. What is the connection between variations in traits and natural selection? (pages 14 and 15)

5. How are artificial selection and natural selection alike? How are they different? (page 16 and 17)

6. Give an example of a predator and its prey and an adaptation each has that helps it survive. (page 19)

7. Use your imagination and write a story about two living things from another planet that depend on each other to survive. Give details about how their symbiosis works. (page 22)

▼ What clever adaptation does this frog have to protect itself from predators? (page 20)

Glossary

adaptation **(n.)** the evolution of features (anatomical, physiological, or behavioral) that makes a group of organisms better suited to live and reproduce in their environment (p. 5)

artificial selection **(n.)** the breeding of selected animals or plants to produce offspring with certain desirable traits (p. 16)

biodiversity **(n.)** the vast diversity of living organisms, their pool of genes, and the ecosystems in which they live (p. 11)

community **(n.)** all of the populations of organisms living in the same environment and interacting with each other (p. 18)

ecological niche **(n.)** the roles and relationships of a particular species in the community of which it is a part (p. 15)

ecosystem **(n.)** all of the living organisms—plants, animals, and microorganisms—that inhabit the same area, interacting with each other and the environment (p. 6)

evolution **(n.)** changes in the collection of genetic material, or the gene pool, from one generation to the next as a consequence of natural selection and other processes (p. 10)

extinction **(n.)** the loss of all individual organisms in a species (p. 8)

fossil **(n.)** the remains or traces of an animal or plant preserved in some way, usually in rocks, but also in ice, peat, or tar (p. 9)

gene **(n.)** the unit of heredity, encoded as a specific segment of DNA, in living organisms. It determines the characteristics that an offspring inherits from its parent or parents. (p. 12)

hibernate **(v.)** to spend cold or dry seasons in a period of energy-saving dormancy and inactivity (p. 21)

microbe **(n.)** (also known as microorganism) a form of life, usually singe-celled, that is too small to be seen without a microscope (p. 21)

migration **(n.)** the seasonal movement of certain animals, mostly birds and fish, to distant places for breeding or feeding (p. 21)

natural selection **(n.)** a process that drives evolution. In natural selection, members of a species with traits that help them survive live and reproduce in their environment, passing those traits on to the next generation. (p. 12)

offspring **(n.)** the immediate descendants of a living organism. Children are the offspring of people. (p. 15)

population **(n.)** all the organisms of one species that live in the same environment at the same time (p. 10)

predator **(n.)** an organism, usually an animal, that captures, kills, and feeds on other organisms (p. 19)

prey **(n.)** the organisms eaten by predators (p. 19)

species **(n.)** a particular kind of living organism. Members of a species have similar characteristics and have the ability to interbreed. (p. 6)

symbiosis **(n.)** a close association between two or more organisms of different species (p. 22)

trait **(n.)** gene-controlled features or characteristics of living organisms that may be anatomical, physiological, or behavioral (p. 12)

Index

adaptations 5, 6, 7, 8, 10, 11, 12, 15, 17, 18, 19, 20, 21, 22, 23, 24, 26, 30
Africa 26, 28, 29
air 5, 18
algae 22
Antarctic 5
ants 23
Arctic ecosystems 7
artificial selection 16, 17, 30
asteroid 8

bacteria 23
bats 24, 25
behavioral ecologist 27
biodiversity 11, 17
blubber 5, 10

camel 7
carbon dioxide 18
cave 24, 25
cells 12, 18, 26
cellular respiration 18
community 18

Darwin, Charles 17
data 25
deserts 7, 19, 21
dinosaurs 8, 9, 11
DNA 12

ecological niche 15
ecosystems 6, 7, 8, 9, 10, 15, 18, 21
  Antarctic 5
  aquarium 6
  Arctic 7

desert 7, 18
  rainforest 7
elephant 26, 27, 28, 29
endangered 24
energy 22, 24
environmental extremes 21
evolution 10, 11, 12, 15, 30
evolve 10, 12, 17
extinctions 8, 9

food 7, 10, 15, 19, 22, 23, 24
fossils 9, 17
fungus 23

generation 12, 15
genes 12, 13, 14, 15

hearing 19
hibernate 21, 24

insects 20, 24, 27
interaction 18

living things 6, 7, 8, 9, 10, 11, 12, 13, 16, 17, 18, 19, 21, 22, 30

microbes 21
microscopic amoebas 19
migrate 21
movement 19

Namibia 27
natural selection 12, 14, 15, 16, 17, 23, 30
naturalist 17
nature 17, 18

O'Connell-Rodwell, Caitlin 27, 28, 29
offspring 15, 16, 17
On the Origin of Species 17
organism 5, 10, 20
oxygen 18, 22

photosynthesis 18
population 10, 13, 14, 15, 18, 25
predators 19, 20, 30
prey 19, 30

reproduce 15
resources 19

savannahs 26
seismic cues 26, 28
species 6, 7, 8, 9, 10, 11, 12, 15, 17, 18, 22, 23
survival 8, 11, 21
survive 6, 7, 8, 9, 14, 15, 20, 21, 22, 23, 24, 26,
symbiosis 22, 23, 30

technology 29
temperature 5, 21, 24, 25
theory of evolution 17
torpor 24
tortoises 14, 15
traits 12, 13, 14, 15, 16, 17, 30

variations 13, 14, 30
vibrations 26

zebras 13

About the Author Rebecca L. Johnson is a national award-winning author of more than 70 books for children and young adults about science. To learn more visit www.SallyRideScience.com.

Photo Credits Rene Krekels/Minden Pictures: Cover. Sebastian Duda: Back cover. Pete Oxford/Minden Pictures: Title page. Eduardo Rivero: p. 2. Mitsuaki Iwago/Minden Pictures: p. 4. René Lorenz: p. 7 top. Andrea Danti: p. 8. Scott Elrick: p. 9 bottom. Jan Daly: p. 10. James Watt: p. 11 top. Jacob Wackerhausen: p. 12. Eric Isselée: p. 13, p. 14 bottom, p. 16 right. Matthew Cole: p. 14 top. Elena Schweitzer: p. 16 left. Layne Gardner: p. 17. James Brey: p. 18. Audrey Snider-Bell: p. 19 top. N. Allin and G.L. Barron: p. 19 bottom. Mads Frederiksen: p. 20 top left. Peter Wey: p. 20 top right. Janet Storey, Carleton University, www.carleton.ca/~kbstorey: p. 21 left. Jon Milnes: p. 22. Ken Lucas/Visuals Unlimited, Inc.: p. 23. Tom Grundy: p. 24. Max Salomon: p. 26, p. 28 right. Courtesy of O'Connell and Rodwell: p. 27, p. 28 left. Courtesy of Stanford University School of Medicine: p. 27 (logo). Ecoimages: p. 29 top. Kitch Bain: p. 29 bottom. Sara Robinson: p. 30.